U.S. Marine Corps War Crimes Investigation

Department of Defense

Pentagon
Publishing

MCRP 4-11.8B

U.S. Marine Corps

PCN 144 000046 00

DEPARTMENT OF THE NAVY
Headquarters United States Marine Corps
Washington, D.C. 20380-1775

22 June 1998

FOREWORD

1. PURPOSE

Marine Corps Reference Publication (MCRP) 4-11.8B, *War Crimes Investigation*, helps Marines determine when specific actions violate the laws of war. This publication provides all Marines with an understanding of their responsibility when non-compliance with or breaches of the law of war are suspected.

2. SCOPE

MCRP 4-11.8B provides specific examples of actions taken by members of a military organization which have been considered noncompliance with or breaches of the law of war. This publication describes the responsibility of each Marine to report suspected war crimes and the reporting procedures. Specific examples of violations of the Uniform Code of Military Justice are given.

3. SUPERSESSION

FMFRP 4-25, *War Crimes Investigation*, dated 3 December 1993.

4. CERTIFICATION

Reviewed and approved this date.

BY DIRECTION OF THE COMMANDANT OF THE MARINE CORPS

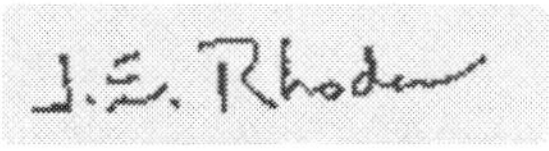

J. E. RHODES
Lieutenant General, U.S. Marine Corps
Commanding General
Marine Corps Combat Development Command

DISTRIBUTION: 144 000046 00

To Our Readers

Changes: Readers of this publication are encouraged to submit suggestions and changes that will improve it. Recommendations may be sent directly to Commanding General, Doctrine Division (C 42), Marine Corps Combat Development Command, 3300 Russell Road, Suite 318A, Quantico, VA 22134-5021 or by fax to 703-784-2917 (DSN 278-2917) or by E-mail to **smb@doctrine div@mccdc**. Recommendations should include the following information:

- Location of change

  Publication number and title
  Current page number
  Paragraph number (if applicable)
  Line number
  Figure or table number (if applicable)

- Nature of change

  Add, delete
  Proposed new text, preferably double-spaced and typewritten

- Justification and/or source of change

Additional copies: A printed copy of this publication may be obtained from Marine Corps Logistics Base, Albany, GA 31704-5001, by following the instructions in MCBul 5600, *Marine Corps Doctrinal Publications Status.* An electronic copy may be obtained from the Doctrine Division, MCCDC, world wide web home page which is found at the following universal reference locator: **http://ismo-www1.quantico.usmc.mil/docdiv**.

Unless otherwise stated, whenever the masculine or feminine

TABLE OF CONTENTS

Chapter I

General Observations

1. Definition

The term "war crime" means a violation of the law of war. Examples of the law of war are the four Geneva Conventions of 1949 which protect wounded, sick, and shipwrecked members of armed forces; enemy prisoners of war (EPW); and civilians who are in your custody. Other examples of the law of war are the Hague Conventions which include rules concerning how we fight (for example, we do not normally select as targets things like churches, schools, and hospitals which have no military value), and what weapons we are allowed to use (for example, we do not use weapons like hollow tip bullets because they are designed to inflict unnecessary suffering).

2. Categories of War Criminals

While most war crimes are committed by members of the armed forces, civilians have been known to violate the rights of individuals protected by the Geneva and Hague Conventions. An example of this is when a civilian attacks or abuses an EPW such as a downed pilot. In this instance the civilian has violated the law of war and that makes him guilty of a war crime. The fact that he is not a member of the military will not save him from being imprisoned or executed if judged to be a war criminal.

3. Serious Violations

While any violation of the law of war is a technical war crime, an investigator should be primarily concerned with the more serious violations which indicate—

- Criminal intent or gross negligence which has resulted in harm (or an attempt to inflict harm) to a noncombatant (for example, sick, wounded, shipwrecked military personnel, EPWs, or civilians who have not become involved in hostilities) and/or—

- An attempt to gain an unfair military advantage by treachery or bad faith (for example, waving a white flag to indicate surrender, and then attacking the individuals who have refrained from firing at you in order to accept a surrender).

4. War Crimes and the Uniform Code of Military Justice (UCMJ)

Although the misconduct to be investigated is a violation of the Geneva or Hague Conventions (which makes it a war crime), the misconduct may also constitute a violation of the UCMJ. For example, the killing by a Marine of civilians in his custody could be characterized as a violation of Article 118 of the UCMJ (Murder), as well as a violation of the Geneva Convention. In most instances, killing a civilian in your custody will be charged as a violation of the UCMJ.

5. Articles 92 and 134 of the UCMJ

Some violations of the law of war are not specifically mentioned in the UCMJ (for example, mutilation of the corpses of enemy soldiers). While this misconduct could be charged as a violation of the law of war, it could also be viewed as a violation of Article 134 of the UCMJ as an offense which involves—

- Disorder or neglect to the prejudice of good order and discipline in the armed forces, or

- Conduct of a nature to bring discredit upon the armed forces.

Violations of the law of war may also be looked upon as a violation of Article 92 of the UCMJ (Failure to obey an order or regulation) because Marines are required to conduct their actions in accordance with the requirements of the law of war.

6. Training and Leadership

Marines are far more unlikely to commit war crimes when they are trained in the prohibitions of war crime activity, when they understand that anyone who commits such activity is a criminal and will be prosecuted, and when they realize that compliance with the prohibitions benefits mission accomplishment. Compliance with the prohibitions of war crime activity not only prevents our most valuable asset (personnel) from being jeopardized, but it also increases public and international support of the U.S. military, encourages reciprocal treatment by the enemy with respect to U.S. Service members who are held captive, and helps end the conflict more quickly by minimizing hostilities. Commanders must personally affirm the prohibitions against war crime activity and impose upon subordinates an expectation that they will not commit war crimes. An explicit and implicit stigma against such activity coupled with a vigilant awareness of factors which contribute to the commission of war crimes can minimize the potential for subordinates to engage in such activity. Historically, the following factors have been found to lead to the commission of war crimes:

- High friendly losses.
- High turnover rate in the chain of command.
- Dehumanization of the enemy or use of derogatory names or epithets.
- Poorly trained or inexperienced troops.
- The lack of a clearly defined enemy.
- Unclear orders.
- High frustration level among the troops.

7. Command Responsibility

Commanders are responsible for training their subordinates on war crimes and investigating war crime allegations. Additionally, commanders are personally responsible for war crimes committed by subordinates if either the commander ordered the commission of the act or the commander knew or should have known of the act, either before or during its commission, and did nothing to prevent or stop it.

Chapter II

Criminal Investigations

1. General

a. Commanders of Marine Corps organizations assigned to unified or specified combatant commands or under the operational control of a commander other than CMC, who receive reports of noncompliance with or breaches of the law of war, shall report the facts to the National Military Command Center in accordance with the directives of their operational commander.

b. All other Marine commanders will report noncompliance with or breaches of the law of war by means of a serious incident report in accordance with MCO 5740.2F.

c. Investigation of suspected violations of the law of war is a command responsibility. Suspected violations of the law of war will be investigated without delay, and reports of investigations and of ultimate disposition will be submitted in accordance with the directives of operational commanders. Violations of the law of war, suspected to have been committed by or against allied military or civilian personnel or against allied military or civilian property, will be reported as directed in (a) and (b) above. However, once it has been determined and reported that U.S. personnel were not involved, additional investigation will continue only by direction of higher authority.

d. Each member of a Marine Corps organization who has knowledge of or receives a report of an apparent violation of the law of war shall as soon thereafter as practical—

- Make the instance known to the immediate commander or commanding officer, or

- If such person has an honest and reasonable belief that the immediate commander or commanding officer is or may be involved in a violation, make the incident known to an officer, normally in the chain of command, senior to the commander or commanding officer.

2. Notoriety of War Crimes

As mentioned above, any violation of the law of war is a war crime; and the person who commits a war crime is a war criminal. Although the military justice system would prefer to treat these violations as UCMJ matters, international public opinion (unfamiliar with the UCMJ) looks upon these matters as war crimes. Due to the notoriety that goes with war crimes, it is especially important for an investigator to perform a diligent investigation. While having a Marine branded as a war criminal is a black eye for the Marine Corps, a slipshod investigation would only add to the bad publicity for the Marine Corps. It is far better for the Marine Corps if a thorough investigation is conducted and appropriate disciplinary action is carried out. The worst possible scenario is for a war crime to be "covered up." In the past, attempted cover ups have served to seriously undermine public support for military operations.

3. Special Considerations

In general, an investigation of an alleged war crime should be treated like any other criminal investigation. However, two considerations should be taken into account:

a. You should anticipate the need for interpreters.

b. Because you will be conducting the investigation in a potentially hostile environment, you may need to request an armed guard force to assure your personal security as well as the protection of your witnesses and evidence.

(reverse blank)

Chapter III

Examples of War Crimes

Listed below are some examples of war crimes. Each of the war crimes described below could be charged as a violation of Articles 92 or 134 of the UCMJ. After some of the examples of war crimes, there appears a list of certain Articles of the UCMJ (in addition to Articles 92 and 134) the violation of which may be charged in connection with the misconduct described in the example.

1. Willful killing, torture, or inhumane treatment of individuals protected by the Geneva Conventions such as—

- EPWs.

- Medical and religious personnel.

- Sick, wounded and shipwrecked combatants (This category also includes crews from disabled aircraft.)

- Civilian internees, refugees and other civilians under a military organization's control.

Inhumane treatment includes willfully causing great suffering or serious injury to body or health. Inhumane treatment may take the form of acts such as beatings. It may also take the form of systematically ignoring or neglecting the needs (shelter, clothing, food, water, and medical attention) of persons entitled to rely upon you to meet those needs under the Geneva Conventions, e.g., EPWs and interned civilians.

An example of mistreatment is ordering an EPW to turn over to his captor his rations, boots, winter coat, first-aid kit and shelter half at a time when the EPW still needs these items.

UCMJ Articles

Art. 93 Cruelty and Maltreatment
Art. 118 Murder
Art. 119 Manslaughter
Art. 120 Rape and Carnal Knowledge
Art. 122 Robbery
Art. 124 Maiming
Art. 128 Assault
General Art. 134 Indecent Assault, Negligent Homicide

2. Extensive unlawful and wanton destruction or illegal taking (confiscation or stealing) of property which is protected by the Geneva and Hague Conventions such as: medical, religious, educational, scientific, and cultural property. This prohibition also concerns private property which has no military value (for example, looting private homes is prohibited). The personal property of EPWs also is protected. Anything a Marine takes from an EPW belongs to the U. S. Government or to the EPW. The Marine has no right to take from an EPW, property which has no military value, such as a wristwatch or money. (NOTE: The taking of "war trophies" [weapons, enemy flags, etc.] is generally not a law of war issue. Service regulations or orders control the items that may be properly taken as war trophies. Marines who violate these regulations or orders may be charged with a violation of Article 92 of the UCMJ. In addition, Congress, at 10 USC 2579, has mandated that all enemy material captured or found abandoned shall be turned into appropriate personnel, as designated by

Service regulations. Subsequent approval of war trophies must come from DOD and the Bureau of Alcohol, Tobacco and Firearms.)

UCMJ Articles

Art. 99	Misbehavior before the Enemy
Art. 103	Captured or Abandoned Property
Art. 109	Property other than Military Property of U.S.—Waste, Spoilage, or Destruction
Art. 121	Larceny and Wrongful Appropriation
Art. 122	Robbery

3. Forcing an EPW or alien civilian to serve in your own forces, or to act as a guide for these forces.

4. Willfully depriving an EPW or a civilian in your custody of the right to a fair and regular trial for offenses allegedly committed by the EPW or civilian. This rule prohibits, among other things, summary executions in the field for alleged misconduct or spying.

5. Unlawful deportation, transfer, or confinement of civilians under the control of a military force or government. For example, when hostilities broke out in Kuwait, U.S. or allied civilians in Kuwait were illegally transferred to Iraq and confined there by the Iraqi Government.

6. Taking innocent civilians and using them as hostages.

7. Using poison or other forbidden weapons such as hollow tip bullets, bacteriological agents, and other weapons determined by higher authority to be illegal because they are calculated to cause unnecessary suffering.

UCMJ Article

Art. 124 Maiming

8. Mistreatment, abuse, neglect, or collective punishment of EPWs (including religious and medical personnel) or civilians in the custody of U.S. armed forces. An example of mistreatment is subjecting an EPW to acts of intimidation, insults, or ridicule. This type of misconduct may take the form of a Marine posing for a photograph of himself holding a gun to the head of a blindfolded EPW.

UCMJ Articles

Art. 93 Cruelty and Maltreatment
Art. 128 Assault

9. Refusal of quarter (refusing to permit an enemy to surrender) unless bad faith is reasonably suspected. You need not stop in the middle of a firefight to accept someone's surrender, nor are you obligated to put yourself at risk in order to accept a surrender. The rule is that you may not refuse to accept a surrender when you have the means to safely accept a surrender.

10. Treacherous (false) request for quarter or misuse of a flag of truce (white flag). This example of a war crime covers the situation in which a Marine attempts to trick his enemy into believing that he no longer wishes to or is able to fight in order to take advantage of the fact his enemy is prohibited from attacking the Marine and is obligated to accept a surrender.

11. Misuse of the Red Cross or other protective emblem. (In most Moslem nations, the protective emblem for

medical service personnel and equipment is the Red Crescent.) An example of this is when a military unit paints a red cross on its building in order to trick the enemy into believing that the building is a hospital which should not be attacked. In addition to the Red Cross, there are protective emblems for religious and cultural buildings. Also, buildings and ships holding EPWs or civilian internees are required to display the emblems "PW" or "IC" which put an enemy on notice that the building or ship contains noncombatants who should not be targeted.

12. Firing on a flag of truce (white flag).

13. Violating surrender or truce terms. For example, attacking the enemy without justification during a truce or a cessation in hostilities.

14. Pillage or purposeless destruction of property.

UCMJ Articles

Art. 99	Misbehavior before the Enemy—Quitting Place of duty to Plunder or Pillage
Art. 103	Dealing in Captured or Abandoned Property
Art. 103	Looting or Pillaging
Art. 109	Property other than Military Property of U.S.—Waste, Spoilage, or Destruction
Art. 121	Larceny and Wrongful Appropriation

15. Theft or looting. (This includes taking personal property from EPWs or civilians for personal enrichment.)

UCMJ Articles

Art. 103	Dealing in Captured or Abandoned Property
Art. 103	Looting or Pillaging

Art. 109 Property other than Military Property of U.S.— Waste, Spoilage, or Destruction
Art. 121 Larceny and Wrongful Appropriation
Art. 122 Robbery

16. Forcing an EPW or civilian in your custody to perform prohibited labor. Ordering EPWs to build a bridge (as in the movie "The Bridge on the River Kwai") designed and constructed primarily for military use is an example of prohibited labor. Labor which is humiliating, inherently dangerous (e.g., clearing mine fields), or hazardous to health is prohibited. Compelling an EPW to perform labor which would help Marines to fight their enemy is prohibited. Ordering an EPW to assist sick or wounded Marines is not prohibited because the Marines are considered to be noncombatants while they are sick or wounded. EPWs may also be required to perform such labor as administrative work, installation repair and maintenance, agricultural work, manufacturing and transportation work which serves no military purpose, domestic work, and work related to the arts.

17. Transferring an EPW to people (military or civilian) of a country which did not sign the Geneva Conventions. For example, if an Israeli pilot were captured by the Iraqis, they could transfer him to Jordan because that nation signed the Geneva Conventions; the Iraqis could not transfer the Israeli pilot (an EPW) to the Palestine Liberation Front.

18. Executing or otherwise punishing (without a trial) an individual accused of being a spy, or a saboteur.

UCMJ Articles

Art. 118 Murder

Art. 119 Manslaughter

19. Attacking an undefended town or a civilian object which has no military significance.

20. Mutilation of corpses (e.g., cutting off ears as a war trophy).

21. Any form of reprisal against a person protected by the Geneva Conventions (for example, a civilian not engaged in hostilities, an EPW, or a sick, wounded, or shipwrecked enemy combatant). If a civilian (an unprivileged combatant) shoots at Marines patrolling an occupied town, he may be fired upon. If the civilian is apprehended, he must be given a trial or hearing before being punished for his actions. If the civilian eludes capture, the Marines may not demolish the home of the parents of the suspected sniper as a form of reprisal. Similarly, if the identity of the sniper is unknown, the Marines may not institute any sort of collective punishment (e.g., cut off electricity or water in the town) directed at the neighbors of the civilian sniper. You may never harm the person or the property of a friend or relative of an offender as a reprisal. Moreover, before reprisals of any kind can be taken they must first be approved by the National Command Authorities.

UCMJ Article

Art. 109 Property other than Military Property of U.S.—Waste, Spoilage, or Destruction

22. Placing a bounty or reward for killing an enemy, individual, or a member(s) of a particular military organization.

23. Attacking Red Cross or Red Crescent personnel, vehicles or buildings, or the wounded they are assisting.

24. Purposefully placing EPWs or civilians in your custody in unsafe or unhealthy locations. For example, forcing an EPW to "take the point" while crossing a suspected mine field.

25. Purposefully placing an EPW or a civilian in your custody (each is considered to be a noncombatant) near legitimate military targets (for example, using them as "human shields"). It is a violation of the law of war for two reasons:

a. If the target is attacked, the EPW or civilian may be harmed.

b. If the attackers refrain from targeting the legitimate military target for fear of harming noncombatants, the Marines who placed the noncombatants in harm's way have taken an unlawful military advantage.

Art. 93	Cruelty and Maltreatment
Art. 118	Murder
Art. 119	Manslaughter
Art. 134	Negligent Homicide

26. Forbidding alien civilians, especially women and children, from departing an enemy nation once hostilities have commenced. (For example, stopping an Iraqi woman living in Saudi Arabia from returning to her homeland.) Note that draft age male civilians may be interned by their enemy if it is suspected that they may enter the armed forces of their nation once they return to their homeland.

27. Failing to notify an EPW's government of the fact the individual has been captured. Also, keeping interned alien civilians incommunicado.

28. Confiscating food, water, medicine, etc. from civilians in your custody due to the fact that they are enemy aliens or are subjects of a country you are occupying. For example, Iraqis confiscated anything of value in Kuwait and shipped it back to Iraq.

UCMJ Articles

Art. 103 Looting or Pillaging
Art. 121 Larceny and Wrongful Appropriation
Art. 122 Robbery

29. Forcing civilians to commingle with troops in an attempt to have your enemy refrain from attacking the combined group.

UCMJ Article

Art. 93 Cruelty and Maltreatment

30. Encouraging or forcing refugees to place themselves in the path of an attacking enemy, in order to hinder his advance.

UCMJ Articles

Art. 118 Murder
Art. 119 Manslaughter
Art. 134 Negligent Homicide

31. When the military situation permits, failure to—

- Search out, collect, make provision for the safety of, or to care for survivors of sunken ships or boats; and

- Care for members of armed forces in the field who are disabled by sickness or wounds, or who have dropped their weapons and surrendered.

32. Intentionally attacking individual civilians who are not actively engaged in direct hostilities against you or the civilian population. This rule prohibits indiscriminate attacks on the civilian population or civilian property, knowing the attack will cause—

- Loss of life or injury to civilians, or

- Damage to civilian property that would be excessive or disproportionate in relation to the concrete and direct military advantage anticipated, and cause death or serious injury to body or health.

UCMJ Articles

Art. 118	Murder
Art. 119	Manslaughter
Art. 128	Assault
Art. 134	Negligent Homicide

33. Poisoning wells, streams, or other water sources.

34. Pretending to be wounded if this is part of a plan to—

- Take advantage of your enemy's obligation not to attack wounded combatants, and

• Attack your enemy when his guard is down. Indicating to an enemy that you are unable due to your wounds to continue to fight has the same effect as indicating (by raising your hands over your head or waving a white flag) a desire to surrender.

35. Forcing an EPW or civilian to collect your wounded fellow Marines from the battlefield during the battle.

36. Physical or mental coercion directed at an EPW or a civilian in your custody to induce him to provide information. For example, threatening to throw someone out of a helicopter unless he provides information.

UCMJ Article

Art. 93 Cruelty and Maltreatment.

20

(reverse blank)

Chapter IV

Incident Report Format

An investigation of an alleged war crime should be reported on a Department of the Navy INCIDENT/COMPLAINT REPORT or a USMC Criminal Investigation Report. If neither of these forms is available, a law of war incident report can take any format as long as the following information is included:

- Narrative description of the incident including: who, what, when, where, why and how the incident occurred;

- Unit and personnel discovering the incident;

- Identity of U.S./enemy units known or suspected of having been in the area at the time of the incident;

- Disposition of the evidence; and

- Any other information that would help an investigator determine what happened and who is responsible for the incident.

WAR CRIMES INVESTIGATION

This booklet was prepared for military police and investigators as a guide in detecting, investigating and reporting war crimes and other violations of the UCMJ.

If you have any questions concerning war crimes, contact the Security and Law Enforcement Branch at Head- quarters, U.S. Marine Corps. The mailing address and telephone numbers are as follows:

COMMANDANT OF THE MARINE CORPS (POS)
Security and Law Enforcement Branch
Operations Division
Headquarters, U.S. Marine Corps
Washington, D.C. 20380-1775

Commercial: (703) 614-4177, 614-2180
DSN: 224-4177, 224-2180

Made in the USA
Monee, IL
04 November 2026

67376403R10020

Praise for *Inside the Minds*

"This series provides a practical and focused discussion of the leading issues in law today." – John V. Biernacki, Partner, Jones Day

"*Inside the Minds* draws from the collective experience of the best professionals. The books are informative from an academic, and, more importantly, practical perspective. I highly recommend them." – Keith M. Aurzada, Partner, Bryan Cave LLP

"Aspatore's *Inside the Minds* series provides practical, cutting edge advice from those with insight into the real world challenges that confront businesses in the global economy." – Michael Bednarek, Partner, Shearman & Sterling LLP

"What to read when you want to be in the know—topical, current, practical, and useful information on areas of the law that everyone is talking about." – Erika L. Morabito, Partner, Patton Boggs LLP

"Some of the best insight around from sources in the know" – Donald R. Kirk, Shareholder, Fowler White Boggs PA

"The *Inside the Minds* series provides a unique window into the strategic thinking of key players in business and law." – John M. Sylvester, Partner, K&L Gates LLP

"Comprehensive analysis and strategies you won't find anywhere else." – Stephen C. Stapleton, Of Counsel, Dykema Gossett PLLC

"The *Inside the Minds* series is a real hands-on, practical resource for cutting edge issues." – Trey Monsour, Partner, Haynes and Boone LLP

"A tremendous resource, amalgamating commentary from leading professionals that is presented in a concise, easy to read format." – Alan H. Aronson, Shareholder, Akerman Senterfitt

"Unique and invaluable opportunity to gain insight into the minds of experienced professionals." – Jura C. Zibas, Partner, Lewis Brisbois Bisgaard & Smith LLP

"A refreshing collection of strategic insights, not dreary commonplaces, from some of the best of the profession." – Roger J. Magnuson, Partner, Dorsey & Whitney LLP

"Provides valuable insights by experienced practitioners into practical and theoretical developments in today's ever-changing legal world." – Elizabeth Gray, Partner, Willkie, Farr & Gallagher LLP

"This series provides invaluable insight into the practical experiences of lawyers in the trenches." – Thomas H. Christopher, Partner, Kilpatrick Stockton LLP

ASPATORE

Aspatore Books, a Thomson Reuters business, exclusively publishes C-Level executives and partners from the world's most respected companies and law firms. Each publication provides professionals of all levels with proven business and legal intelligence from industry insiders—direct and unfiltered insight from those who know it best. Aspatore Books is committed to publishing an innovative line of business and legal titles that lay forth principles and offer insights that can have a direct financial impact on the reader's business objectives.

Each chapter in the *Inside the Minds* series offers thought leadership and expert analysis on an industry, profession, or topic, providing a future-oriented perspective and proven strategies for success. Each author has been selected based on their experience and C-Level standing within the business and legal communities. *Inside the Minds* was conceived to give a first-hand look into the leading minds of top business executives and lawyers worldwide, presenting an unprecedented collection of views on various industries and professions.